Habitats

Making Homes for Animals and Plants

BY Pamela M. Hickman

ILLUSTRATED BY Sarah Jane English

**FEDERATION OF
Ontario
Naturalists**

Addison-Wesley Publishing Company
Reading, Massachusetts   Menlo Park, California   New York
Don Mills, Ontario   Wokingham, England   Amsterdam
Bonn   Sydney   Singapore   Tokyo   Madrid   San Juan   Paris
Seoul   Milan   Mexico City   Taipei

In memory of my parents, Marguerite and Melville Hunter – PMH

Library of Congress Cataloging-in-Publication-Data

Hickman, Pamela M.
  Habitats : making homes for animals and plants / by Pamela M.
Hickman ; illustrated by Sarah Jane English.
    p.   cm.
  Includes index.
  Summary: Includes directions for creating temporary habitats
indoors that allow for the study of various animals and plants in
natural settings.
  ISBN 0-201-62618-7
  1. Habitat (Ecology)—Experiments—Juvenile Literature.
2. Ecology—Study and teaching—Activity programs—Juvenile
literature. [1. Animals—Habitat. 2. Plants—Habitat.
3. Ecology—Experiments. 4. Experiments.]
I. English, Sarah Jane, 1956-   ill. II. Title.
QH541.24.H53 1993
574.5'078—dc20   93-12683
                                                            CIP
                                                            AC

Many of the designations used by manufacturers and sellers to
distinguish their products are claimed as trademarks. Where those
designations appear in this book and Addison-Wesley was aware of
a trademark claim, the designations have been printed in initial
capital letters (e.g., Popsicle).

Neither the Publisher nor the Author shall be liable for any damage
which may be caused or sustained as a result of conducting any of
the activities in this book without specifically following instructions,
conducting the activities without proper supervision, or ignoring the
cautions contained in the book.

The Federation of Ontario Naturalists (FON) is a non-profit
membership organization that is committed to protecting and
increasing awareness of important habitats, wilderness areas and
endangered wildlife. The FON cooperates with national and
international organizations to achieve these goals. For more
information on the FON, contact that Federation of Ontario
Naturalists, 355 Lesmill Road, Don Mills, Ontario M3B 2W8,
Canada.

Other nature-activity books by Pamela M. Hickman and the FON:
*Birdwise: 40 Fun Feats for Finding Out About Our Feathered
Friends*
*Bugwise: 30 Incredible Insect Investigations (and Arachnid
Activities)*

Edited by Laurie Wark
Designed by Michael Solomon
Typeset   by Esperança Melo
Printed and bound by Metropole Litho Inc.

 Text stock contains over 50% recycled paper

1 2 3 4 5 6 7 8 9—959493
First printing, August 1993

Contents

A Note to Parents and Teachers

The natural world is an unlimited source of beauty and fascination. Whether it's an anthill in a sidewalk crack or a moss-covered tree in a forest, there are always new discoveries in every outing. The ideas in this book will supplement a child's outdoor experiences and add an exciting intimacy with the natural world. These activities encourage children to take a closer look at nature by creating temporary mini habitats at home or in school. By viewing the growth, natural changes, life cycles and special adaptations of plants and animals, children can observe and understand many aspects of nature that are impossible to see in the field. With this book, nature can become a year-round hobby and a lifelong adventure.

Part of the learning experience gained through these activities involves caring for wild plants and animals and ensuring their safety and survival. Read the Conservation Tips on the next page before you begin the activities. The mini habitats created are clearly temporary, and children are encouraged to carefully return everything to its original habitat after their observations are over. This process develops respect for other life and helps ensure that the natural environment is disturbed as little as possible. The protection of our natural heritage is vitally important. One way to help ensure the future of our wildlife and wild areas is to involve children in learning about the wonders of nature. With knowledge and fascination come respect and a sense of responsibility for the environment.

Conservation tips

Here are a few ways you can protect nature while you're learning about it.

1. When you are outdoors, observe as much as you can in its natural setting. Bringing a plant or animal home temporarily can be a terrific opportunity to learn more, but it is not wise to try this with everything. For example, vertebrates (animals with backbones), especially mammals, are difficult to keep, and they will suffer if they are confined in a small space. Likewise, many plants, such as orchids, are very sensitive to disturbance and often don't survive transplanting. Most states and schoolboards have laws against keeping wildlife, so you should check out your local laws before proceeding.

2. Always get permission from landowners before removing anything from the wild. Never take anything from a protected area such as a park, conservation area or nature reserve.

3. Take only a few individual plants or animals from any area, and don't take home more than you can safely care for. The container for your temporary habitat should be prepared before you collect the wild plants or animals.

4. Any species that are uncommon, rare or endangered should never be collected under any circumstances. Contact your local conservation or wildlife officer for information about which species are at risk.

5. If a plant or animal does not appear to be thriving, return it to its original habitat as soon as possible. All organisms must be returned to the wild when you have finished observing them.

What's a Habitat?

Look around your home. It provides you with shelter, food and a place to live and play. Your home is your habitat, the way an anthill is the home, or habitat, for an ant. And just as your home is part of a larger town or city, the ant's habitat is part of a larger community called an ecosystem. Many different habitats make up an ecosystem such as a field or forest where thousands of plants and animals may live. With this book you'll discover what goes on in a variety of habitats by getting nose-to-nose with some of nature's fascinating creatures. By creating mini habitats in jars and aquariums, you can check out a few days, weeks or months in the lives of some amazing plants and animals. You'll find out why worms stretch and shrink, how plants recycle air and water, how tadpoles become toads and much more. These things happen in nature all the time, but we don't often get to see them close up. By bringing a bit of nature indoors, you'll see things happen right before your eyes.

Taking care of a piece of nature is lots of fun, but it's also a big responsibility. You must make sure that all the plants and animals you collect are well cared for. This book tells you what you'll need to know to look after your mini habitats safely. However, if your mini habitat isn't doing well, return it to the outdoors. All the creatures should be returned to their original homes once you've finished watching them. When you release or replant them, the organisms can continue on with their lives and with their role in the habitat. Your mini habitats are small parts of

much bigger ecosystems. All these ecosystems are connected to one another, and they all work together to keep the world healthy. Just like people, plants and animals depend on their habitat for survival. Right now, some natural habitats are not being well cared for. Garbage is dumped in fields, chemicals such as phosphates are poured into rivers and lakes, and acid rain is hurting the soil and water in many habitats. When a healthy habitat becomes polluted, it is no longer a good home for all its plants and animals. And as we build more roads, houses and factories, some plants and animals are losing their wilderness homes and they are becoming endangered and extinct — gone forever. We're even losing entire habitats — rainforests are being burned, old-growth forests are being logged and wetlands are being drained. Once you've watched and cared for wild plants and animals, you'll discover how important it is to respect and protect their natural habitats. Making the mini habitats in this book is a great way to find out more about habitats and how you can help protect them.

Most of the habitats in this book can be made either in a 4-L (1-gallon) jar or in an aquarium. The container that works best for each habitat is mentioned first in the list of materials you'll need.

7

Life in a Rotting Log

When you're walking in the woods, look around for trees that have fallen to the ground. They may have been blown down in wind storms or broken by other falling trees. These dead trees become a small community of plants and animals. You can bring a piece of a fallen tree home and put it in a terrarium where you'll discover this amazing community of living creatures for yourself.

You'll need:
a small aquarium or a 4-L (1-gallon) jar (a huge condiment jar works well)
tape
2 sticks or long pencils (for jar only)
small pebbles or sand
small, untreated charcoal pieces (available at grocery or hardware stores)
part of a rotting log
a trowel
soil from a forest floor, including the decaying leaves on top
water
woodland plants and fungi from around the log
fine screening
an elastic (for jar only)

1. Place a layer of small pebbles or sand on the bottom of your terrarium for drainage. If you are using a jar, turn it on its side. Tape the two sticks or pencils to the jar to keep it from rolling.

2. Sprinkle a thin layer of charcoal pieces over the pebbles. This will help to keep the soil fresh.

3. If possible, take your terrarium to the woods. You can also collect all the materials in containers or bags and assemble them in your terrarium at home. Make sure that live animals, such as insects, are carried in containers with air holes punched in the lids.

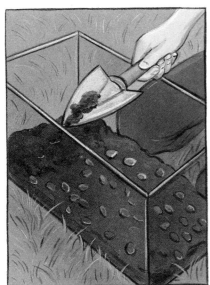

4. Find a small rotting log, full of life. With the trowel, put about 7 cm (3 inches) of nearby soil and decaying leaves into the terrarium. Add water, if necessary, to make the soil mixture damp. Shape the soil into small hills so that it is more like the forest floor.

5. Carefully break off a small piece of the rotting log. Put the piece of log, and 2 cm (3/4 inch) of the soil from right beneath the log, into the terrarium. Your piece of log should fit in your terrarium without being squished.

6. Dig up a few of the mushroom-like fungi and small plants from around your log and plant them in your terrarium. Press the plants down firmly and water them. Your terrarium should be kept moist but not soaked.

7. Cover the terrarium's opening with a piece of fine screening and secure it with tape or an elastic.

8. Place your terrarium in a north-facing window where it will get natural light, but not direct sunlight. It also needs fresh air and temperatures between 18°C and 24°C (65°F and 75°F). Keep it away from heaters and drafts.

9. Watch your mini log habitat for signs of life and discover how the fungi, plants and animals will help to turn your log into soil. You can keep your terrarium going for several months — it makes a great winter hobby. After you've finished observing your terrarium, return everything to its original home, either before late fall or after the ground has thawed in the spring.

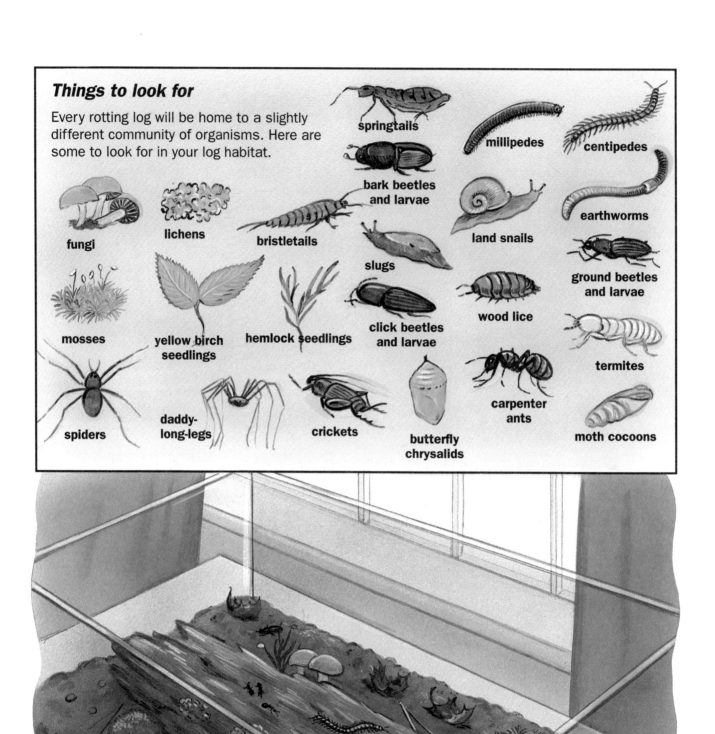

Things to look for

Every rotting log will be home to a slightly different community of organisms. Here are some to look for in your log habitat.

springtails

millipedes

centipedes

bark beetles and larvae

earthworms

fungi

lichens

bristletails

land snails

slugs

ground beetles and larvae

mosses

yellow birch seedlings

hemlock seedlings

click beetles and larvae

wood lice

termites

spiders

daddy-long-legs

crickets

butterfly chrysalids

carpenter ants

moth cocoons

Habitat Watch

Your rotting log is like a mini community. Just as the people and places in your community change over time, the animals and plants that live in the log community change, too. While some die out or move on, others stay and new ones move in. What you find in your piece of rotting log will depend on what kind of tree it is, how long the log has lain on the forest floor, the season when you find it, where the tree grew and the climate in which it grew. Your terrarium contains just a moment out of the entire life of a rotting log community. To get a broader picture, you can set up several terrariums at once. In one terrarium, place a newly fallen log, in another put a log that has started to decay, and in a third terrarium get a log that is starting to fall apart because it is so rotten. You can compare the different fungi, plants and animals in each log and see for yourself how the log community changes over time.

As your log slowly rots, the wood is being broken down into tiny bits that will eventually become part of the soil. Turning a log into soil is a lot of work and it takes a

team effort. The first helpers — fungi and bacteria — are so small that you need a microscope to see them. They soften up the wood with special chemicals so that larger animals, such as insects, can eat it.

Once the wood has been worked over by the fungi and bacteria, new helpers join the rotting team. Look for the grayish-brown common wood lice (for more on wood lice, see page 20). Insects also invade the log, searching for food and shelter. Lift up a piece of bark. You may find an artistic-looking pattern of tunnels and chambers. The artists of these patterns are bark beetles, sometimes called engraver beetles. The adult beetles arrive in the log first, drilling out tiny bedrooms where their eggs are laid. When the young insects hatch, they create tunnels by eating their way through the wood. Do you notice that their tunnels get wider as they go along? As the young eat more wood, they get fatter and have to make wider tunnels to fit through.

You might also find large, black carpenter ants tunnelling into the wood and making large holes for nesting. As all these creatures tunnel and burrow into the wood, they leave many spaces for fungi to grow deeper into the log, decaying it from inside as well as outside. With more and more holes appearing, water can also seep farther into the wood. Moisture and warm temperatures make the log rot faster.

In the forest, the log becomes a hive of activity, with small creatures climbing in, out, over and under it. Your log won't attract toads, salamanders, mice and snakes to your terrarium, but in the forest these larger animals come to rotting logs for many reasons. Hollows in the rotting wood make cosy, sheltered places to rest, lay eggs and raise a family. The log itself is like a supermarket full of food for these hungry bug catchers.

Animals aren't the only log invaders. The velvety carpets of mosses and the crusty or leafy growths of lichens also belong to the log community. Even trees, such as hemlock or yellow birch seedlings, may take root in the rotting wood. These trees have a hard time growing through the thick layer of leaves on the forest floor. So seeds that are lucky enough to land on rotting logs get to start life well above ground and get a boost up.

When the log is completely rotten, it eventually falls apart and everything tumbles down to be added to the forest floor. Look at the bottom of your log where it seems to become part of the soil. You are seeing the final result of the work of thousands of organisms over many years. The rotted wood is now used as a fertilizer by trees and other plants to help them grow. When they die, the whole cycle begins again.

THE BIG PICTURE

As a tree grows, it takes nutrients from the soil to help make its wood. When the tree dies and rots, the nutrients it once used to help it grow are returned to the soil. This soil helps new trees to grow and so on. If fallen trees are not left to rot but are cut down and taken away, the nutrients aren't returned to the soil and it becomes poorer. In some parts of North America, such as British Columbia and Oregon, large areas of land are stripped of every tree during logging. In many countries in Central and South America and Africa, areas of rainforest are being cleared for farming. Although the soil is rich at first, it soon runs out of nutrients since it is not being recycled through the process of rotting and growth. In a few years, the soil cannot grow anything and the area is abandoned. When the cycle of life and decay is broken, it doesn't take long for a forest or jungle to turn into a wasteland.

Wonderful Worms

Think of an animal that has no eyes, no ears and no legs. Need another hint? It's also one of the hardest workers under the ground. If you guessed an earthworm, you're right. Digging in the garden is one way to see earthworms, but building a wormery is an easier way to watch these incredible soil makers at work.

You'll need:
a 4-L (1-gallon) jar (a huge condiment jar works well) or a small aquarium
some damp soil
some damp sand
dead leaves
lettuce, spinach or other leafy vegetables
some worms
a trowel
a flashlight (optional)
a plastic container with holes punched in the lid (a yogurt container works well)
cheesecloth or thin gauze
an elastic
tape
black construction paper

1. With the jar standing up, place a layer of damp soil, about 4 cm (1 1/2 inches) deep, in the bottom of your jar. Add a layer of damp sand, about 4 cm (1 1/2 inches) deep. Continue to add alternating layers of soil and sand until the jar is nearly full. Keep the jar upright.

2. Place some dead leaves and a few bits of lettuce on top of the soil and sand for food.

3. Collect a few worms by digging them up in a garden, moist field or woods. During the day you might have to try worm calling (see the box on the next page). At night take a flashlight and you should find worms on top of damp soil. Gather three or four worms and put them in a plastic container with some damp soil and a few rotting leaves. At home, place the worms in your wormery.

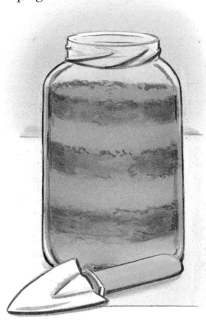

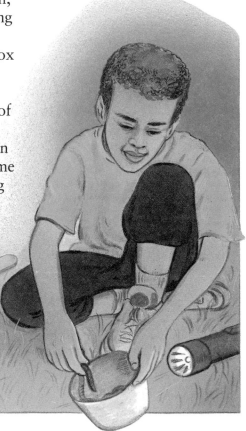

4. Cover the jar's opening with a piece of cheesecloth secured with an elastic. If you are using an aquarium, tape the cloth in place.

5. Tape a piece of black construction paper around the jar to encourage the worms to make their burrows near the glass.

6. Keep your wormery in a shady, cool place, such as a basement or back porch, and keep the soil damp but not water-logged. Feed the worms with bits of leaves and lettuce every few days.

7. After a week, remove the black paper for a short time to see how the wormery has changed.

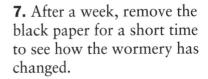

Worm Calling

You've heard of whistling for your dog, but what about calling for worms? Since worms are nocturnal (active at night), during the day you may have to call them out of the ground with this simple trick. Hammer a wooden stake into the ground and then rub a board over the top of the stake. When the board rubs the stake, it sends vibrations into the ground and signals danger to the worms below. Once a worm feels the vibrations, its first reaction is to leave its tunnel as quickly as possible. You'll soon see worms popping out of holes in the ground.

8. You can keep your wormery for many months and watch it all winter, if you want. Remember to feed your worms every few days. Collect some leaves in plastic bags in the fall to add to the wormery over the winter. When you are finished watching your worms, return the worms to their original home in spring, summer or early fall.

Habitat Watch

Check on your worms for a little while every day. Are they more active in the morning or afternoon? Can you see any worms pushing their way through the soil? Worms make tunnels by eating through the soil and swallowing earth and the bits of plants and animals that get in their way. Eventually, the digested food comes out the other end of the worm as worm castings. Look for these tiny, tidy piles of pellets left on top of the soil. Do the worms live close together or are they usually alone? If you see two worms lying side-by-side with their heads at opposite ends, they're probably mating.

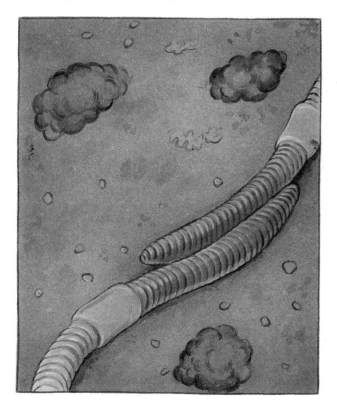

Can you tell which end of an earthworm is the head and which is the tail? Watching a worm move isn't a reliable clue since it will move both backward and forward. Look for the smooth, light brown, un-lined part of the worm's body. This is called the clitellum, and it is usually closest to the worm's head. The head end is also slightly more pointed than the tail end.

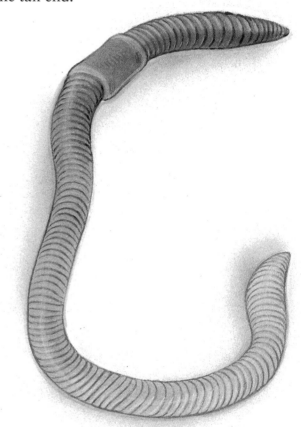

What does a worm feel like when you pick it up? The slimy, wet feeling comes from a layer of mucus on the worm's skin. Worms don't have lungs inside their bodies to help them breathe, like you do. Instead, they breathe right through their skin. Since oxygen is absorbed more easily through wet skin, worms must stay damp to survive. The skin acts not only as a worm's lungs, but also as its eyes. A worm's skin is sensitive to bright light, and earthworms usually avoid it because too much light can dry out their skin.

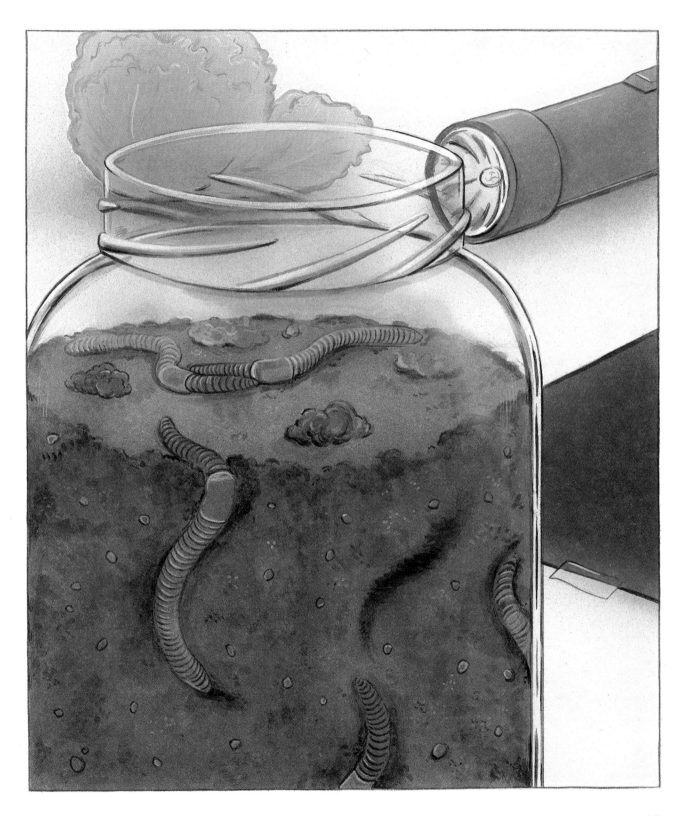

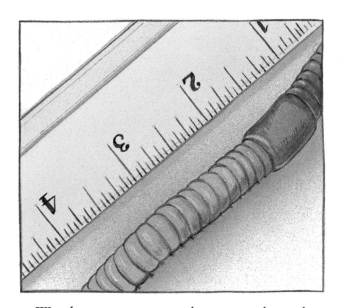

Watch your worms as they move through the soil. What do you notice? The incredible shrinking worm? Place a small ruler next to a moving worm and measure the worm's body length at different stages in its motion. The worm's body is covered with tiny lines that divide it up into little parts called segments. As a worm moves, the segments are stretched apart and then pushed back together. It's like a Slinky toy that gets longer and shorter as you pull it out and push it back together. On most segments, there are eight small bristles to help the worm move. Try to find them with your magnifying glass. Place a small piece of paper in your wormery and put a worm on it. Listen for the scratching noises made by the bristles.

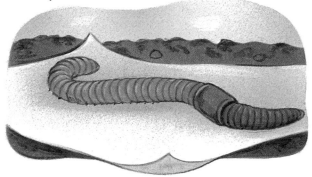

Try this

• Try shining a red light over your wormery in a darkened room. Don't put the light too close to the jar or it will dry out the soil. Do the worms avoid the red light or do they just ignore it? Worms  can't see red light so they don't notice it.

• Are your earthworms picky eaters? Find out what their sense of taste is like by performing a simple taste test. Cut small squares from different leaves; include  things such as lettuce and spinach, as well as maple, oak, cedar or other plants in your area. Remove all the food from the wormery and then spread your buffet of leaf samples on the soil. The next day, check to see which leaves were eaten and which were ignored. Like most animals, earthworms eat what tastes good to them and ignore what they don't like.

• Earthworms are also known for their keen sense of touch. Try this experiment that was first performed more than 100 years ago by Charles Darwin, the famous biologist. Place your wormery on top of a piano. What happens when you play high notes and low notes? Allow a few minutes between notes. The vibrations from the low notes will travel into the wormery and send a danger signal to the worms, who will leave their burrows (see *Worm Calling, page 15*). The high notes vibrate very little and may have no effect on the worms.

What did you see when you took the black paper off your wormery? Could you still see the layers of soil and sand? Your worms were busy making their burrows in the dark, and they mixed up the soil and sand layers and even dragged some of the surface leaves underground to line their burrows and to eat. A farmer plows fields to loosen the soil, mix up the nutrients and allow air and water to get into the soil to help the plants grow. Earthworms do all these things, too, without any machinery. An earthworm's tunnels create air spaces in the soil, allowing air and water to reach down deeper into the soil, where plant roots grow. Worm castings create a nutrient-rich fertilizer for the soil that also helps plants grow. So wherever earthworms live, they help the soil and make the habitat of nearby plants a better place to live.

Watching Wood Lice

Dark, damp and dirty. That's the perfect habitat for wood lice. Whether they live in fields or forests, wood lice are important helpers in their habitat, recycling dead plants and sharing the food with others. With some simple materials, you can make a terrific wood lice farm and find out more about these little land-loving cousins of lobsters and crayfish.

You'll need:
a 4-L (1-gallon) jar (a huge condiment jar works well) or a small aquarium
tape (for jar only)
2 sticks or long pencils (for jar only)
a measuring cup
some damp compost or garden soil
damp peat moss
sand
rotting leaves
some bark
a small piece of sponge
water
some wood lice
a plastic container with holes punched in the lid (a yogurt container works well)
fine screening
an elastic

1. Turn the jar on its side and tape two sticks or pencils to it, as shown, to keep it from rolling around.

2. Mix 500 ml (2 cups) of damp compost or garden soil with 250 ml (1 cup) of damp peat moss and 250 ml (1 cup) of sand. Add this mixture to your jar.

3. Place rotting leaves over the soil mixture for food and add a couple of pieces of bark for shelter from heat and light. Put in a small piece of soaked sponge for additional moisture.

4. Look for wood lice in damp places under logs, stones, old boards and rotting leaves, or make a wood lice trap (see the box on page 21). When you find wood lice, collect them in a plastic container with some rotting leaves and take them home to your terrarium.

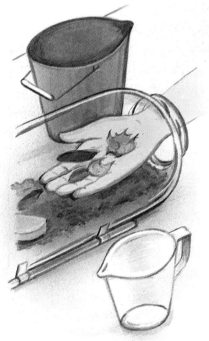

5. Cover the jar's opening with fine screening and secure it with an elastic or tape.

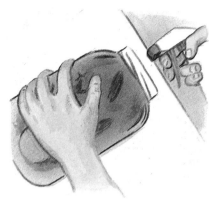

6. Keep your terrarium damp by spraying it lightly with water, when necessary. Put it in a shady area such as a basement or back porch.

7. You can watch your wood lice for several weeks and then return them to their original homes before winter.

Make a wood lice trap

Potatoes are good to eat but they also make great wood lice traps. Use a spoon to scoop out the middle of a few raw potatoes and place them outside in shady, damp areas. The wood lice are attracted to the dark, damp hollows in the potatoes. After a few hours, or overnight, check your potato traps for wood lice.

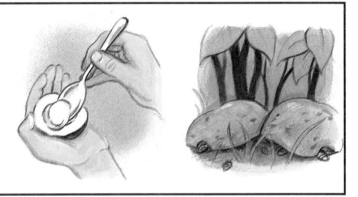

Habitat Watch

When you grow, your clothes get too small and you have to get new ones. It's the same with a wood louse. When its armor gets too tight, it splits from side to side across the middle and is shed. This is called molting. The new, larger armor grows underneath the old covering and is light-colored, soft and stretchy at first. Look at your wood lice to see if any of them are half dark-colored and half light-colored. These are in the middle of molting. In a few days to two weeks, the molt will be completed.

Wood lice produce eggs but they never need a nest for them. The female carries her eggs in a pouch hanging from her belly until the baby wood lice are ready to hatch. Try to find a wood louse with an armor-covered pouch hanging from its belly. Keep a close watch on the mother so you can see when the babies hatch. How do the babies and adults differ? You will be able to spot the babies because they are smaller than the adults, are paler in color and are born with only six pairs of legs, instead of seven. They will grow another pair of legs later. The young stay with the mother for a few days and then wander off.

Your wood lice probably all look the same, but you may have two different kinds — sowbugs and pillbugs. You can tell them apart by watching how they defend themselves from danger. Pick up some wood lice in your hand. What do they do? Pillbugs curl up into a tiny ball when they're frightened, protecting their soft belly from enemies while leaving only their tough armor exposed. The armadillo, a large armored mammal from South America, does the same thing. Sowbugs don't curl up when they're frightened. When they're attacked by hungry predators, such as birds and spiders, sowbugs release a sticky, smelly liquid to get rid of enemies.

Did you know that the pillbug's scientific name — Armadillidium — comes from its resemblance to the armadillo?

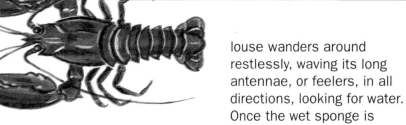

Try this

Believe it or not, wood lice are closely related to lobsters and crayfish. Wood lice live on land, but they breathe through gills as fish do. Wood lice live in damp areas to stay moist because their gills work only when they're wet. Since wood lice need a moist habitat to survive, they are experts at finding water. Test this out with a simple experiment. Put a wood louse in a small box of dry sand for a few minutes. How does it behave? Now place a water-soaked piece of sponge in one corner of the box. What does the wood louse do? In the dry sand, the wood louse wanders around restlessly, waving its long antennae, or feelers, in all directions, looking for water. Once the wet sponge is added, the wood louse immediately rushes over to it and settles down.

THE BIG PICTURE

Good neighbors in your community help each other out. In the soil community, wood lice are good neighbors to have. They play an important part in breaking down large bits of rotting plants into small, bite-size pieces for millions of tiny soil animals that share their habitat. These animals rely on the small bits of food left by wood lice to survive.

Slugs and Snails

Lift a large stone or an old board from some damp ground and you'll probably find a soft, slimy, blob-like slug or a coiled snail underneath. Can you guess why slugs are slimy or imagine what a snail's tongue looks like? Did you know that snails can make their eyes disappear? You can discover much more about these amazing animals by bringing home some slugs and snails and raising them in a terrarium.

1. Place a 1-cm (1/2-inch) layer of sand in the bottom of the aquarium or jar. Cover this with at least 8 cm (3 1/4 inches) of damp soil. If you are using a jar, turn it on its side and tape two sticks or pencils to it to keep it from rolling.

2. Carefully plant some mosses and a few small plants in the soil. Add a piece of bark and some large stones to give the slugs and snails somewhere to shelter from the heat and light. Snails also need a piece of chalk or limestone to nibble at to help their shells grow.

You'll need:
a small aquarium or a 4-L (1-gallon) jar (a huge condiment jar works well)
tape
2 sticks or long pencils (for jar only)
sand
damp soil
moss and a few other small plants, such as grass
a piece of bark
some stones
a small piece of chalk or limestone
some land snails and slugs
a trowel
a spoon
a plastic container with holes punched in the lid (a yogurt container works well)
fine screening
an elastic (for jar only)
water
dark-colored construction paper
food for slugs and snails: green plants, such as leaves or shredded lettuce or cabbage; dead leaves; chopped apples; carrots or potatoes; bread crumbs; and cheese or bits of raw meat

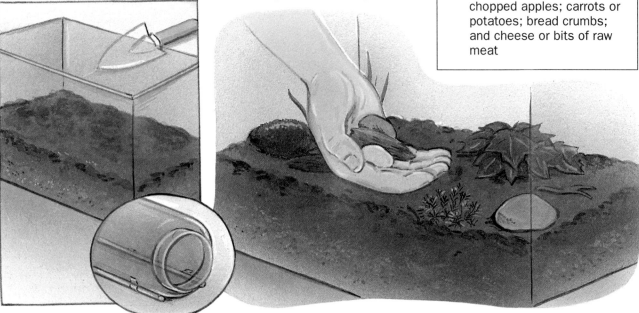

3. To collect some slugs and snails for your terrarium, look under large stones, bark, rotting leaves, old boards or fallen logs in damp places such as gardens, woods or under a porch. Look for the soft, slimy, blob-like slugs and the coiled shells of land snails. The tiny, round pearl-like eggs of both slugs and snails may be found in the same places, as well as several centimetres (a few inches) below ground in loose, damp soil.

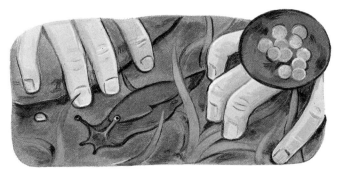

4. Gently pick up a few slugs, snails or eggs with your fingers or a spoon and place them in a small container with some damp soil.

Take them home and transfer them to your terrarium. The eggs should be buried just below the soil surface or under bark or stones where it is dark. If you cover the outside of the terrarium with dark paper (see Step 7), you can place the eggs on top of the soil, near the glass, where you can watch how they change.

5. Cover the terrarium's opening with screening and secure it with tape or an elastic, if you're using a jar.

6. Keep the terrarium in the shade and sprinkle it with water every few days to keep the soil and air moist.

7. Cover the lower outside part of the terrarium with dark paper. This may encourage the animals to lay eggs against the glass. You can observe them by removing the paper for short periods of time.

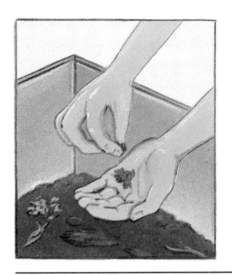

8. Feed the slugs and snails small amounts of food every day, and clear away uneaten food twice a week.

9. You can keep your slugs and snails for several weeks or months, or even all winter. When you have finished watching the slugs and snails, return them to their original home when the ground is warm in spring, summer or fall.

Habitat Watch

What's the difference between a slug and a snail? Slugs and snails are closely related. On the outside, slugs and snails are easy to tell apart: a snail's soft body is covered by a hard, coiled shell, but a slug's body has no shell. On the inside, however, slugs and snails are very similar, and they move, eat and breathe alike.

How can you tell a male snail or slug from a female one? You can't. That's because each snail or slug is both a male and a female at the same time. When an animal contains both male and female sex organs, it is called a hermaphrodite. It still takes two snails or slugs to mate, but both will produce eggs. If you watch your animals on a warm night, you may get to see them do their courtship

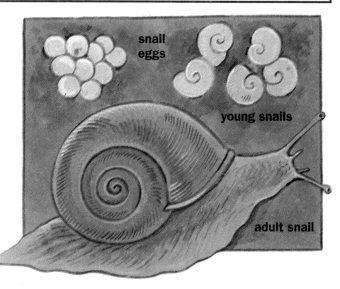

snail eggs
young snails
adult snail

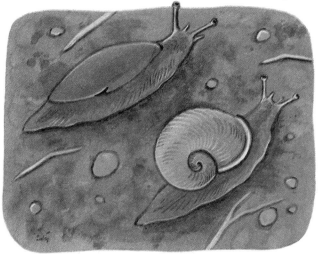

"dance." Watch as they circle each other slowly, come close together and rise up until the bottoms of their feet are pressed together. After clinging together for several hours, they separate. Up to 50 eggs each are laid after a week or two. They take three to four weeks to hatch. After hatching, the babies eat their eggshells and climb to the surface to feed on plants or soil. Notice the shells of the baby snails. How many whorls can you count? Baby snails have soft and see-through shells with only two whorls instead of the four or five whorls on an adult's shell. A baby snail grows by adding coils to its shell. Baby slugs don't have shells. Their bodies get longer and thicker as they grow.

Try this

• Imagine having your teeth attached to your tongue. Snails and slugs have a tiny, roughly toothed tongue called a radula. Use a magnifying glass and try to see their tongues while they are feeding. Instead of taking bites out of the food, the snail uses the radula like a file, rasping back and forth over the food and shredding it into little bits. The way the tongue moves is also a clue to a snail or slug's identity. Scientists can tell different kinds of slugs and snails apart by studying their tongue prints. It's a bit like checking people's fingerprints. You can collect tongue prints from your snails and slugs with this easy activity.

You'll need:
beef fat
a jar with a lid
garden soil
a slug or snail

1. Ask an adult to melt the beef fat and pour it into the jar lid. Let it cool until the fat hardens.

2. Put some soil and a slug or snail in the jar and *loosely* screw the lid onto the jar. Leave the jar in a cool, shady spot overnight.

3. Early the next morning, check the lid for tiny patterns. These are the radula patterns of your snail or slug. Is there more than one pattern? To compare different patterns, you can place different slugs and snails in separate jars overnight and repeat the experiment.

• It's not surprising that snails and slugs move so slowly. After all, they have only one foot. Watch how your snails and slugs move. Place a strip of black construction paper across the soil in your terrarium so that the animals will cross it. What do you see on the paper after they have crossed? Both slugs and snails produce a slimy liquid called mucus. The mucus keeps the animals moist and lays a slippery path to glide on. If you pick up a slug or snail gently and let it crawl on your hand, you will be able to feel and see its slime trail on you. Don't worry about getting slimed — it's harmless.

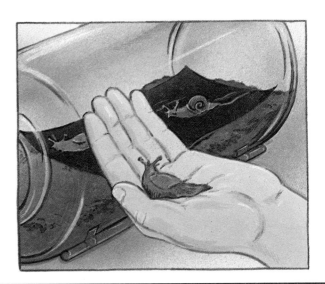

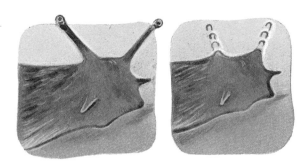

• Use food as bait to observe your snails and slugs in motion. Place food in one corner of the terrarium and watch how the animals behave. Two pairs of tentacles help a slug or snail explore its surroundings. The smallest pair of tentacles is used for smelling and feeling. The largest tentacles have eyes on the ends and are used for seeing. What happens when you bring an object close to their eyes? The tentacles can turn inside-out, making the eyes disappear.

• What do you think snails and slugs prefer to walk on? Find out by setting up a simple experiment. Place several pieces of the same food in the terrarium and surround each piece with a different substance, such as sand, grass, ash, gravel or earth. Which pieces of food are eaten first? Do the animals avoid any of the substances? Although their slime helps them slide over sharp or rough surfaces, slugs and snails don't like the feel of very dry or coarse textures and usually stay away from them. Some gardeners protect their plants from hungry slugs and snails by spreading sand or ash around the base of each plant.

THE BIG PICTURE

Snails and slugs are an important link in the food chain of their habitat. Most snails and slugs are herbivores, which means they eat plants. Their sharp tongues help shred plants into tiny pieces that smaller soil animals can eat, too. Some snails and slugs are carnivores, or meat eaters, that feed on live insects, earthworms or other slugs and snails. When slugs and snails eat dead plants and animals, they help recycle the nutrients in the habitat. Many animals, including birds, shrews, moles, mice, frogs and toads, depend on slugs and snails as their food.

Look for these body parts on your slugs and snails:

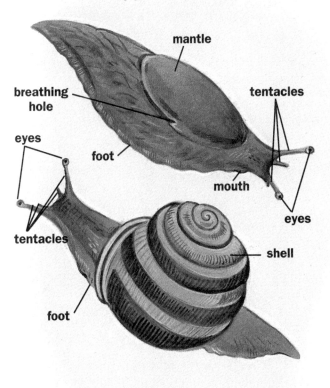

mantle

breathing hole

tentacles

eyes

foot

mouth

eyes

tentacles

shell

foot

Did you know that some kinds of slugs and snails are as small as this dot · , and others are as long as this page?

29

Ponder a Pond

Ponds are great places for people to swim and skate, and they're also terrific habitats for plants and animals who find food and shelter there. Wildlife from nearby habitats depend on ponds, too, for water and food. Fish, birds, turtles and frogs are some of the biggest food in a pond, but the most important pond animals are the smallest ones. Without pond insects and other tiny water creatures, the larger pond animals would starve. You can see how pond insects live by setting up your own mini pond in an aquarium. Just borrow a few things from the kitchen and head to your nearest pond.

You'll need:
a kitchen strainer tied to a broom handle for a longer reach
some large plastic pails with lids (4-L or 1-gallon ice-cream pails work well)
aquatic plants, insects and other invertebrates
rubber gloves (optional — good in cold water)
a field guide to insects and pond life (optional)
plastic bags and twist ties
a shallow light-colored dishpan
a turkey baster
small plastic containers with holes punched in the lids
a small aquarium
small stones
tweezers
fine screening
tape

1. Using the strainer, scoop up enough muck from the bottom of the pond to fill your aquarium about 3 cm (1 inch) deep. Place it in one of your pails.

2. Choose a few underwater plants, such as *Elodea* (check your field guide), and place them in plastic bags filled with pond water and tied with twist ties.

3. Collect enough clear pond water in a pail to fill your aquarium three-quarters full.

4. Fill your dishpan with clear pond water and use it to hold the insects and other animals you catch.

5. Dip the strainer into the water to catch swimming insects. Jiggle the strainer gently along plant stems and under floating leaves. Try to catch any creatures that fall free and put them in the dishpan.

6. Collect one or two pond snails from plant leaves or stones to put in the aquarium. The snails help keep the water clean by eating the dead plant and animal materials and the tiny bits of green algae that will grow on the sides of your aquarium.

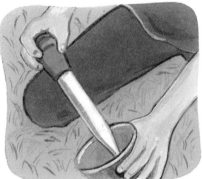

7. When you have several different kinds of creatures, use the turkey baster or tweezers to carefully transfer them to your small plastic containers with some pond water and aquatic plants. Put the lids on, making sure there are air holes punched in the lids.

8. Aquatic insects that live on the surface, such as water striders, should not be transported in water since they may drown as the water sloshes back and forth. Put them in a container with damp aquatic plants only.

9. At home, set up your aquarium in a bright area, but not in direct sunlight (a north-facing window is good). Add the muck to the bottom and carefully put the plants in, holding the roots in place with small stones. Slowly pour the pond water into the aquarium until it is three-quarters full. Wait until the water has cleared before adding the pond insects.

10. Use tweezers to carefully transfer your animals to their new home. Cover the container with screening secured with tape.

11. When you've watched your aquarium for one or two weeks, return all the plants and animals to their pond while the weather is still warm. If you want to keep your pond habitat for longer than a couple of weeks, you'll need to replace half the water with fresh pond water every two weeks to provide a new supply of plankton to feed the tiny aquatic animals. As some of your insects are eaten, or turn into adults and are

released, you can replace them with new insects from the pond. Your mini habitat will require continual additions and replacements, or the predators will starve once their food supply is eaten. In a natural pond, new animals are always being born, so food is always available. Return your pond creatures to their original habitat before the first frost in fall so they have lots of time to prepare for winter.

Habitat Watch

Do any of your creatures change while they are in your aquarium? Over a couple of weeks, you may see young insects (larvae or pupae) develop into adults right before your eyes. Put a long stick in the aquarium for dragonfly, damselfly and caddisfly nymphs so they can climb out of the water to transform into adults. Let the winged adults go once you've had a good look at them.

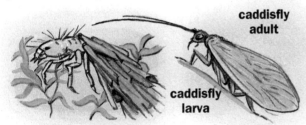

caddisfly adult

caddisfly larva

Do you see any creatures hanging just below the surface? Mosquito larvae, called wrigglers, hang upside-down from the water's surface and poke a tiny, tube-like siphon into the air. The siphon acts like a snorkel, drawing air into the mosquito's body. Mosquito pupae, which look like tiny, black commas, have a pair of siphons on their head, like little antennae.

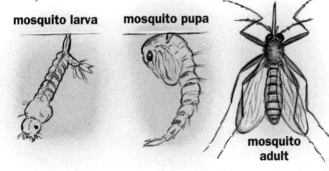

mosquito larva   mosquito pupa

mosquito adult

You may notice some small, fast-moving, black beetles that swim in circles on the surface and then dive down underwater. These are whirligig beetles. Look for a tiny bubble of air at the tip of their abdomen as they dive down. Is it still there when they come back up? This bubble, along with a bubble that is trapped under their hard wing covers, acts like a scuba tank, storing oxygen to breathe while the beetles are underwater. When the oxygen is all used up, the beetles resurface for another "tank" of air.

whirligig beetles

Can you see any creatures that don't come to the surface for air? These animals may have gills that let them take oxygen right out of the water, as fish do. Damselfly nymphs have three large, tail-like gills at the ends of their bodies. Very tiny animals breathe right through their skin, so they can stay underwater all the time.

How do you think the submerged plants are "breathing"? Look for tiny bubbles on the underwater leaves of your plants. As a plant undergoes photosynthesis to make its food, called glucose, it takes carbon dioxide out of the water and gives back oxygen. The bubbles you see are oxygen. This oxygen is used by the other pond life for breathing. The animals, in turn, breathe out carbon dioxide for the plants to use. It's a perfect partnership!

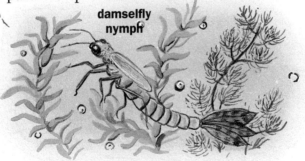

damselfly nymph

Peek at plankton

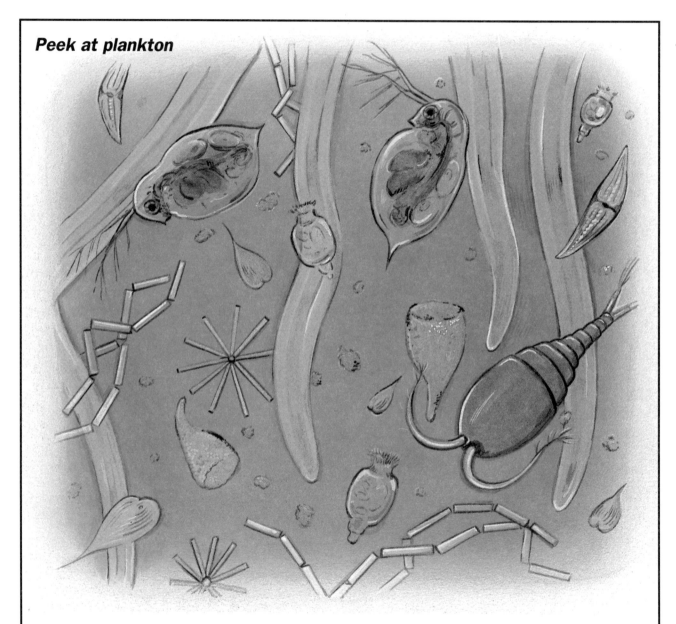

Did you know that there's a whole community of plants and animals in your aquarium that are so tiny you can't see them without a microscope? The plankton community is full of life in beautiful shapes, sizes and colors. If you have a microscope, use an eyedropper to place a drop of pond water on a slide and look at it under the lens. In general, the life you see moving around is animals, and the life with green coloring is plants. How many different kinds can you see? Plankton may be tiny, but they are a very important part of life in a pond. They provide most of the food for insects and other small invertebrates, which in turn feed the frogs, toads, snakes, turtles, fish, birds and mammals of the pond.

Life in the pond is an insect-eat-insect world. For instance, dragonfly nymphs eat mosquito larvae. Can you figure out which animals are the hunters and which are the prey in your mini pond? The animals rely on each other for food and use the plants for food and shelter, too. Their lives depend on having a healthy and clean pond habitat in which to live, just the way people need a healthy and clean environment. Can you tell if your pond water is clean or polluted? The color of the water is not a reliable clue since some pollution is invisible. One way biologists discover whether a pond is healthy is by looking at the animals that live there. Some animals are found only in very clean water, while a few do well in highly polluted water. The kinds of animals that scientists look for are called indicator species; they tell us, or indicate, if the water is clean or polluted. Check the mini chart below to see if your pond has any of these indicator species.

When a pond habitat is polluted with chemicals, detergents, oil or other things, it changes. Many of the plants and animals that live in the pond will die or will become sick and won't be able to produce any more babies. When this happens, the natural balance in the habitat is upset. The food for some animals may disappear, so they must move away or starve. If the plant eaters leave, some plants may spread too quickly and take over the habitat, spoiling the homes of fish and other animals that need open water.

Polluted water can also harm the animals from nearby habitats that come to a pond to drink and eat. Some pollution can stay in the mucky pond bottom for years, slowly poisoning the water and its plants and animals. Although polluted habitats can sometimes be cleaned up, they will never be just the same as they were before the balance was upset.

Your pond is clean if you find these animals:

Your pond is polluted if you find mostly these animals:

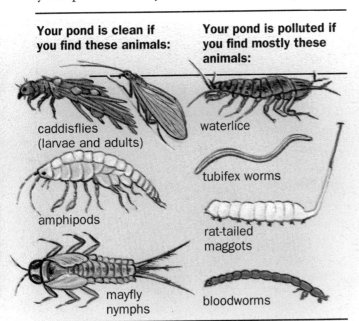

caddisflies (larvae and adults)

amphipods

mayfly nymphs

waterlice

tubifex worms

rat-tailed maggots

bloodworms

A Toad's Tale

When you were a baby, you looked like a small version of your parents. Baby amphibians (animals that live on land and in water), such as toads and frogs, don't look anything like their parents. When they hatch from their eggs, toads and frogs are tiny, fish-like tadpoles. Create a mini habitat to raise some toad tadpoles indoors and discover how these tiny swimmers turn into land-loving hoppers.

You'll need:
toad eggs
a kitchen strainer
a large plastic container
    with lid (a 4-L or 1-gallon
    ice-cream pail works well)
pond water
food for tadpoles, such as
    lettuce and hard-boiled egg
pieces of bark
clean coarse gravel or sand
a 20-L (5-gallon) aquarium
bits of charcoal
soil
a shallow bowl
small rocks and stones
twigs
tape
fine screening
live food for adult toads such
    as mealworms (available
    from pet stores),
    earthworms (see page 15),
    beetles, moths, earwigs
raw hamburger and thread
    (optional)

1. In the spring you'll find toad eggs in small ponds that dry up during summer and in shallow areas of larger ponds. Look for long strings of jelly-covered eggs wrapped around underwater rocks, twigs or plants. Don't confuse them with the floating masses of jelly-covered frog eggs.

2. Use your hands or a kitchen strainer to scoop up some eggs and place them in a large plastic container with about 7 cm (3 inches) of pond water. Put the lid on your container and take it home.

3. At home, remove the lid and place the container in a bright area out of direct sunlight.

4. After 3 to 12 days, the eggs should hatch, depending on the water temperature; the warmer the water, the faster they'll hatch. If there are a lot of tadpoles in the small space of your aquarium, the larger ones will eat the smaller ones, so keep only about 4 tadpoles and return the rest to the pond.

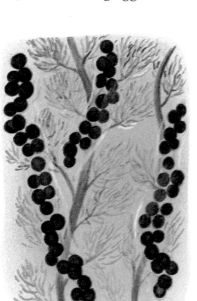

5. Tadpoles eat the algae (tiny green plants) in the pond water so change half the water twice a week to keep a fresh supply of algae in the container. Simply scoop out half of the old water with a cup and carefully pour in the new pond water. You'll need to go back to the pond where you found the toad eggs to get the fresh pond water. You can also feed tadpoles small pieces of boiled or wilted lettuce or bits of hard-boiled egg every day. Clean out any uneaten food with a strainer before adding more.

6. When the tadpoles develop legs, float some pieces of bark in the water so they can climb out of the water. When their tails disappear, you need to transfer the toads to a terrarium. Keep only one or two toads and return the rest to the edge of the pond where you found the eggs.

7. To make a terrarium, place a 5-cm (2-inch) layer of clean gravel or sand in the bottom of a 20-L (5-gallon) aquarium tank. Cover this layer with charcoal pieces to keep the terrarium fresh, and add 10 cm (4 inches) of soil. The soil should be kept damp but not soaking.

8. Place a shallow bowl of pond water in the soil at one end of the terrarium. Add a few small stones to the water to give the toad a grip on the smooth dish. Change the pond water twice a week.

9. Place some small rocks, twigs and bark over the soil to provide shelter for the toads. The toads will dig in the soil and uproot plants so it is best not to plant anything.

10. Add the toads to the terrarium and tape a screened cover in place to keep them from jumping out. Place the terrarium in a north-facing window away from direct sunlight, where it will get lots of fresh air, but not be in drafts.

11. Feed the growing toads daily with live food. You can raise your own food, catch it outdoors or buy it at a pet store. Some toads will also eat bits of raw hamburger if it is placed on the end of a thread and dangled in front of the toad, as if the food were alive. If your toad is ignoring its food, there may be something wrong. Return the toad to its pond where it will be able to feed and grow naturally.

12. After a week of toad watching, return the toads to the edge of the pond where you found the eggs.

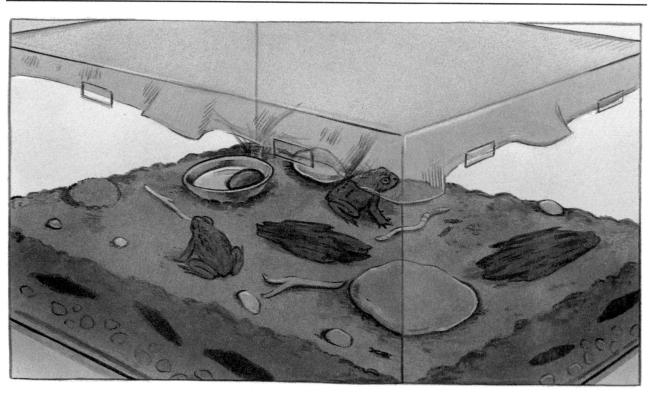

Metamorphosis Watch

Get ready for one of nature's best magic acts. Before your eyes you'll see eggs hatch into tadpoles and then transform into adult toads. This change is called metamorphosis. While you are raising your eggs and tadpoles, make notes about what you see and draw pictures or take photographs so you'll be able to use them as a science project or tell your friends about it. Look for these signs of change.

1. The jelly surrounding the eggs protects the eggs from damage and keeps them afloat. When the eggs hatch, tiny black tadpoles rise up in the jelly and cling to it for a day or so. At first the tadpoles have very small tails and are weak swimmers so they use the jelly as a life raft. They stay together in groups until they are large enough to swim on their own. Watch how their tails wiggle continuously.

2. Use a magnifying glass to see the tiny gills at the sides of their heads. Tadpoles breathe through gills, as fish do.

3. In three to four weeks you'll notice tiny bumps on both sides of a tadpole's body, near the base of its

1

2

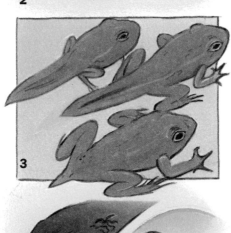

3

4

5

tail. Watch over the next week or so as these bumps gradually grow into short hind legs. About two weeks after the hind legs grow, the front legs appear out of the gill openings on each side of the tadpole's body. Which leg comes out first? What happens to the tadpole's tail? As a tadpole grows larger, its tail shrinks until it eventually disappears.

4. The shape of a tadpole's mouth changes, too. The small, round mouth of a tadpole has sharp jaws and tiny teeth for scraping algae and protozoans (microscopic animals) off rocks and plants. The mouth gradually widens into a large, toothless mouth that makes a toad look as if it has a grin on its face. Inside its body, the toad has replaced its gills with lungs for breathing air and has also changed its digestive system because adults eat meat instead of plants.

5. In about two months, when the tadpole's tail disappears, it is ready to live on land. The tiny black or dark brown toad is only a few centimetres (1 inch) long and looks like a cricket when it moves onto the land.

Habitat Watch

Can you find your toad's ears? Look for the smooth, round shapes on the sides of its head, behind the eyes. These are its eardrums, which pick up sound vibrations, like your own eardrums. Toads can hear very well, and that's especially important during mating season when the males sing loudly to attract females. Does your toad have a baggy, white-

skinned pouch at its throat? This is a male toad's vocal sac. You can hear the springtime chorus of male toads in woodland ponds.

The thick, rough skin of a toad feels dry compared to the slippery skin of a frog. A toad's thick skin doesn't lose water as easily as a frog's, so toads can spend a lot of time on land, away from a pond. Toads absorb moisture through their skin instead of drinking through their mouths. If a pool of

water is not nearby to lie in, toads can soak up the morning dew or burrow into damp ground. As the tiny toad grows, its skin gets too small, and it must grow a new one and shed the old skin. A young toad grows so quickly that it sheds its skin every week. The whole process takes only five minutes, so you'll have to watch carefully to see it. The toad's old skin splits along its back and the toad pulls it off, like you'd pull off a sweater. You won't find the old skin lying around, though, because the toad eats it!

You have to be really fast to catch a fly with your hand, but imagine being able to catch flies with your tongue! Watch your toad catch its dinner. How close does the food get to the toad's mouth before it is eaten? Your tongue is attached near the back of your mouth, but a toad's tongue is attached to the front of its mouth. When a toad flicks its tongue out, the

back of the tongue shoots forward so it can reach way out. Passing bugs get caught on the sticky tongue and are pulled back into the toad's mouth. Are your toad's eyes open or closed when it eats? Toads always eat with their eyes closed. That's because the toad's eyeballs sink into its head and help push the food down its throat. Since a toad swallows quickly, your toad will look like it's blinking when it eats.

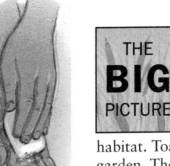

If you've heard that touching toads gives you warts, don't worry — because it isn't true; the bumps on a toad's back are not warts. The two largest bumps behind a toad's eyes are actually poison glands used to keep enemies from eating the toad. It's a good idea to wash your hands after touching a toad since the poison can irritate your mouth or eyes.

Try this

• Toads have several tricks for avoiding enemies such as snakes, birds and raccoons. A toad's brown, rough skin blends in well with the ground so it's hard for enemies to see it. If an enemy does catch a toad, the toad's poison glands release a very bad-tasting liquid so that animals will spit it out. Snakes don't seem to be bothered by the poison, however, so toads puff up their bodies with air to make themselves too big for a snake to swallow. Try putting a rubber snake in the terrarium with your toad to see if it will puff itself up in defense.

THE **BIG** PICTURE

Tadpoles are an important source of food for fish, turtles and other pond animals. As adults, toads eat thousands of insects, helping to maintain the balance within their land habitat. Toads are also a big help to your garden. They catch lots of bugs that eat plants and pester people. In fact, some pioneers used to keep a toad in the house to control the bugs indoors, too.

Although toads spend a lot of time on land, they can lay eggs only in water. As people drain or fill in ponds and marsh habitats for farming or construction, toads have fewer places in which to lay eggs. This means that fewer toads develop each year. Toads are great at defending themselves against most enemies, but they can do nothing to protect themselves against the destruction of their habitat. Only people can stop that.

Amazing Ants

An ant colony is like a miniature palace where servant ants run around attending to the queen and looking after all the other household chores. In this hidden underground habitat, ants work, rest, eat and sleep safely. You can make an anterrarium and find out what ants do when they disappear underground, how they send messages and why they're called social insects. You may even get to meet the queen.

You'll need:
a 4-L (1-gallon) jar (a huge condiment jar works well) or a small aquarium
an empty frozen-juice can with one end sealed, or a larger container, if using an aquarium
some soft garden soil
a small piece of sponge soaked in water
food for ants, such as bread crumbs, sugar or honey
a trowel
some ants
cheesecloth or thin gauze
an elastic band
tape
black paper

1. Place the juice can, open-side-down, in the middle of the jar. Fill the space between the sides of the jar and the can with loosely packed soil, leaving 5 to 7 cm (2 or 3 inches) of space at the top. The juice can forces the ants closer to the glass where you can see them better. If you're using an aquarium for your ant colony, put a larger container, such as a plastic food saver, in the aquarium instead of a juice can.

2. Put a small piece of soaked sponge in the jar for moisture.

3. Find an anthill of small black or brown ants in an area that will be easy to dig in, such as a backyard or garden. Sprinkle some sugar and bread crumbs over the hole to attract the ants. Dig in a wide circle around the hole and scoop as much of the anthill as possible into your jar. Try to find the biggest ant, the queen, and put her in your jar along with as many other ants as you can get from the same hill.

4. Cover the jar's opening with cheesecloth and secure it with an elastic band.

5. Tape a piece of black paper around the jar up to the level of the soil. This will encourage the ants to tunnel next to the glass. After about a week, remove the paper for a short time to look at the tunnels and ants underground.

6. Keep the soil moist, but not wet, and place the terrarium in a shady place that has good air circulation. Feed your ants every day by sprinkling some honey-covered bread crumbs and sugar over the soil. Remove any leftover food before adding more.

7. Watch your ants for a while every couple of days and try some of the investigations on the following pages. You can keep your anterrarium for many months, as long as you continue to feed the ants and make sure they are thriving. An ant colony can be fun to keep over the winter, too. When you've finished with your anterrarium, return the ants to where you found them in spring, summer or fall.

What if...

...you find winged ants in your colony? These are the new queens and males, or drones, and they are produced once a year. Release the winged ants outdoors and let them fly off to mate.

...the colony gets too crowded? The more workers you have in your jar, the faster your colony will grow. When it is too crowded, the queen stops laying eggs. If this happens, return your ants to their original habitat, or release half of the colony, not including the queen, and watch the colony rebuild itself.

Habitat Watch

Your habitat is your home, the place where you live. The ants' habitat is their underground colony. If you had to move house as quickly as the ants in your anterrarium just did, it would take a long time to organize your things and get everything back to normal. Ants are masters at organization, though, and they can build a new colony within days, complete with bedrooms, pantries for food and even special rooms for their garbage. While the queen rests in the throne room laying eggs and being groomed, all the worker ants are busy doing housekeeping, maintenance, nursery duties, grooming and feeding each other and collecting food.

You'll see adult ants, but you'll also see ant eggs, larvae and pupae. These are the stages in an ant's life cycle. The eggs are tiny and white. After about a week in her new home, the queen should begin laying eggs again. Look for workers carrying them away from the queen. The tiny larvae look like wriggling worms and the pupae are similar to little grains of rice. When you find some larvae and pupae, watch how the workers busily feed the larvae and wash and groom them along with the pupae.

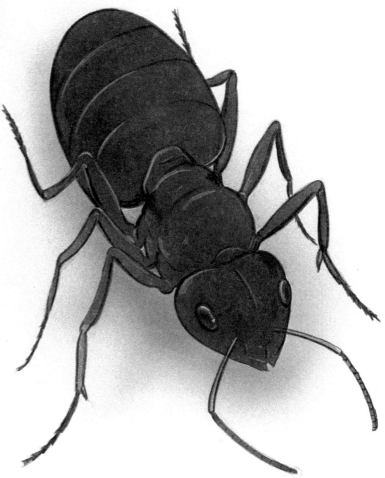

If you were blindfolded, you would likely wave your arms in front of you to help you feel your way through a room. Ants do this with their antennae. Since ants have poor eyesight, they constantly twitch their antennae to feel their way around. Look for the elbow-like joint in the middle of an antenna. It helps the antenna bend and move all around. Antennae are also important for smelling, picking up vibrations and even taking the temperature. Use a magnifying glass to watch an ant grooming its antennae. Try to find the built-in "comb" near the middle joint of each front leg. It's used for cleaning the antennae and the rest of the body.

Try this

• Have you ever wondered how so many ants manage to discover your sandwiches at a picnic? You can find out how ants spread their news by sprinkling some sugar-coated bread crumbs in one corner of your anterrarium. After the first ant finds the food, follow it as it goes back to the colony to spread the news. How does it communicate with the ants it meets? As they rub together or lick each other, the first ant is passing taste and smell messages to the others, explaining that she has found food. Watch how the other ants travel to the food. Do they follow the first ant's trail, or go different ways? What happens if you rub your finger in a line across the path the ants are following? When the first ant walked the trail, she let her abdomen drop down to the ground several times along the way, leaving a scent trail for the others to follow. When you rub your finger across the trail, you erase the scent at that point. The ants are confused for a while but eventually the right trail is found again and they continue on their way.

• Unlike most insects, ants are social insects. This means that they work together to help one another survive. Watch how the worker ants co-operate to share the task of carrying large bits of food back to the colony. Find out how big a load your ants can carry by sprinkling different-sized bits of bread on the soil. Some ants can lift up to 50 times their own weight. That's like you lifting two small cars at the same time!

• Ants need warm temperatures to stay active. When cold weather comes, the workers slow down until they eventually freeze to death in winter. Check out how cool temperatures affect your ant colony's activities by placing your anterrarium in the fridge (*not* the freezer) overnight. The next morning, remove it from the fridge and watch how the ants behave. The workers will move very slowly at first, but they will gradually become more active as they warm up.

• Poke a long, thin stick down one side of the anterrarium to make a hole. Fill the hole with water to represent a flood in the colony. What do the ants do? You should see them moving eggs and young to a drier, safer section of the colony. How long does it take for the ants to repair the damage?

Ants may be a nuisance at picnics, but they're a great help to the plants that share their habitat. As they tunnel through the ground, ants loosen up the soil and make it easier for plant roots to spread. The space created in the soil also allows air and water to reach farther down into the ground, helping plants grow. Their non-stop activity in the soil makes ants terrific soil mixers and movers, too. Instead of using a shovel or rototiller to turn the soil in your garden, ants do it for you just by travelling around. In fact, ants living in half a hectare (1 acre) of land can move a dumptruck-full of soil a year! Keeping the soil well mixed brings the nutrients from the surface down to the plant roots where they are needed.

A Forest under Glass

You'll need:
gravel
a glass aquarium
charcoal pieces
woodland plants
potting soil
a trowel
plastic bags
small rocks or pieces of bark
a spoon
water
a plant mister
a lid or plastic wrap to cover
   the container

When you walk through a forest, you can see and hear lots of different animals moving around, but do you know what the plants are doing? You may not see any action, but the plants are all busy making food, breathing and recycling water. To get a close-up view of all this activity, plant a mini forest habitat and discover how nature uses the air and water over and over again to help the forest grow.

1. Place a layer of gravel 2 or 3 cm (about an inch) deep in the bottom of the aquarium for drainage.

2. Add a layer of charcoal pieces to keep the soil fresh and place 5 cm (2 inches) of potting soil on top of the charcoal.

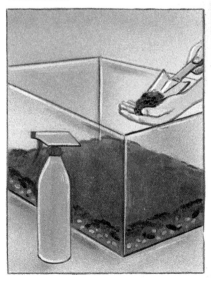

3. Before you dig up some woodland plants, make sure you have permission. You cannot take anything from protected areas such as parks or conservation areas. The plants you choose should be small and grow slowly (see **What to grow**, page 51). Use a trowel to dig up the plants; place them in plastic bags, along with a bit of soil. Disturb the area as little as possible and take only a few plants.

4. Collect a few small rocks or pieces of bark, preferably with moss or lichen growing on them. These can be added to the container later.

5. When placing your plants in the aquarium, arrange them in a forest-like scene. Use taller plants to represent trees and shrubs, and ground covers, ivies and mosses can cover the soil. Make sure you leave lots of space for the plants to grow. If you want to view your mini forest from all sides, place large plants in the center and shorter plants around the edges. Dig small holes with a spoon, put the plants in and press the soil down around the roots.

6. When all your plants are in place, add the small rocks and bark. Water the container with a plant mister until the soil is damp, but not soggy.

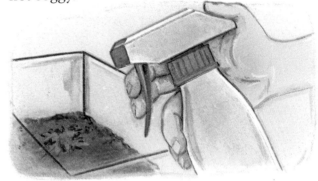

7. Cover your container and place it in a bright area, such as a north-facing window, but not in direct sunlight. Some woodland plants, such as ferns, grow best in semi-shade.

8. Your mini forest looks after itself. Every month check to make sure the soil is still damp, since you may need to add a bit of water occasionally.

9. After a year or so, you may need to replace some plants because they are too big, or not very healthy anymore. You may also want to redesign your scene and add different kinds of plants. Replace the old soil with fresh soil to give the plants a new supply of nutrients.

10. You can keep your mini forest for many years with some annual pruning or replanting. When you have finished with the plants in your forest under glass, replant them in their original habitat.

What if...

...the glass is all fogged up? This means there is too much moisture in the container. Take the cover off and let the container dry out for a few days and then put the cover back on. If you see condensation (water droplets on the inside of the glass) on the top third of the container only, this means that the moisture level is okay.

...the leaves on your plants are turning brown? Your plants may prefer less heat and sunlight, so try putting the container in a more shaded area.

...a plant gets too tall for the container? Pinch off the tops of the stems with your fingers. This will make the plant send out new branches farther down the stem and slow down the plant's upward growth. Plants that grow too quickly and begin to crowd out other plants should be removed and replaced with smaller plants.

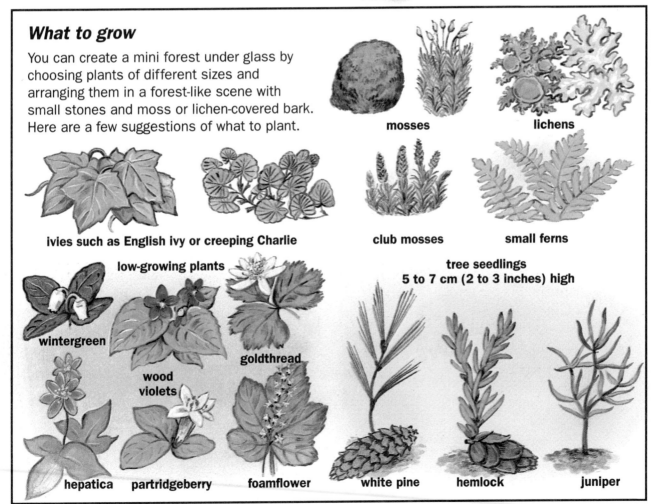

What to grow

You can create a mini forest under glass by choosing plants of different sizes and arranging them in a forest-like scene with small stones and moss or lichen-covered bark. Here are a few suggestions of what to plant.

mosses

lichens

ivies such as English ivy or creeping Charlie

club mosses

small ferns

low-growing plants

tree seedlings 5 to 7 cm (2 to 3 inches) high

wintergreen

goldthread

wood violets

hepatica   **partridgeberry**   **foamflower**   **white pine**   **hemlock**   **juniper**

Habitat Watch

Have you ever seen a fountain of water in a park or shopping mall? Water sprays out of the fountain into a shallow pool below. Then that water is pumped back up inside the

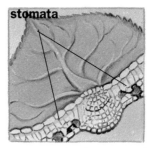

fountain so it can spray again. The same water goes up and down all day. The plants in your terrarium are like water fountains. The roots suck water up from the soil and send it up the stem to the leaves. The leaves get rid of most of the water through tiny holes called stomata — this process is called transpiration. Look for tiny droplets of water collecting at the edges of your plants' leaves. These droplets of water are turned into water vapor (an invisible gas) and are evaporated by the heat of the sun. In a forest, water vapor rises up into the sky, condenses (turns back into water droplets) and forms clouds. The water vapor can't go through the glass of your container, so when it reaches the top of the container, it cools down and condenses on the glass. The water drips down the sides of the aquarium to the soil where it is ready to be sucked up by the plants' roots again. Just like the fountain, your plants use the same water over and over again.

You need good food to help you grow, but what makes plants grow? Instead of finding

food to eat like animals do, plants make their own food in their "leaf factories." The green leaves of plants contain chlorophyll, which takes energy from the sun, carbon dioxide from the air, and water from the soil to make food for the plant. This process is called photosynthesis. Watch for new leaves, branches or stems appearing on your plants. Use a ruler to measure the height or length of your plants' main stems every month. Which plants grow the fastest?

Unlike water that is recycled by the plants, the soil minerals stay in the plant. The minerals are not returned to the soil until the plant dies and decomposes. As your plants continue to grow, they use up all the minerals in

the soil. To keep your plants healthy, every year you need to add fresh soil that has a new supply of minerals.

The oxygen you're breathing right now may have been made by your houseplant or a neighbor's tree! During photosynthesis, plants take in carbon dioxide and give out oxygen. The oxygen is used by you and other animals for breathing. In turn, you breathe out carbon dioxide for plants to use. Since there are no animals in your closed aquarium, why don't the plants run out of carbon dioxide? In addition to photosynthesis, plants also breathe, or respire, especially in the dark when their leaf factories shut down. Respiration is the opposite of photosynthesis; plants take in oxygen and give off carbon dioxide. The plants in your container are recycling the air all by themselves.

When you put the cover on your container, you create a closed ecosystem. This means that nothing is added to or escapes from the container. Our whole world is a type of closed ecosystem, too, since air and water are never lost or added — they are continually recycled. If our air and water become too dirty from pollution, there are no new supplies to use.

If you have trees in your neighborhood, then your air is being cleaned for free. Trees and other plants absorb large amounts of carbon dioxide and dust particles from the air, making it cleaner and healthier for us to breathe. The loss of our forests means that less carbon dioxide and other pollution is being taken out of the air. Too much carbon dioxide in the air contributes to the greenhouse effect. Carbon dioxide acts like panes of glass in a greenhouse by trapping heat and reflecting it back to Earth. Scientists are worried that the greenhouse effect may cause Earth to get warmer. This global warming could cause many problems, including rising sea levels and coastal flooding as polar ice-caps melt. Many cities are planting trees to help clean the air and absorb the carbon dioxide made by automobiles and factories.

When a forest habitat is destroyed, the plants and animals that live there are lost, too. But the effects of the habitat destruction can also be much greater. When large areas of forests are cleared, the water cycle is also harmed. Instead of rainwater being absorbed by the tree roots as usual, it runs off the land into rivers, lakes or other waterways. Without the trees, less water is put back into the air and the cycle is broken. Over many years, the loss of plant life can cause droughts in some areas or create deserts where almost nothing can grow.

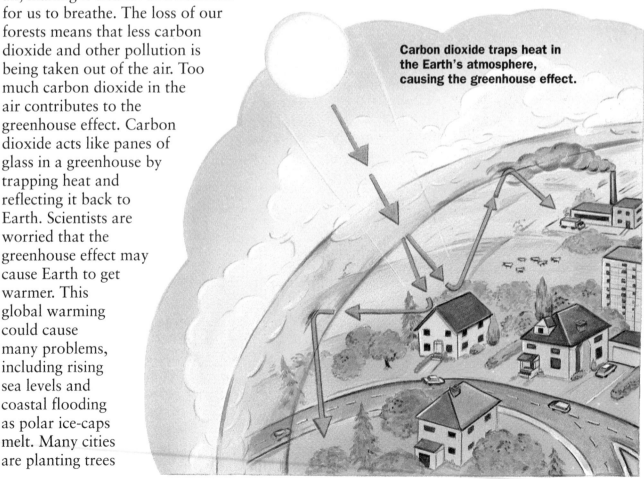

Carbon dioxide traps heat in the Earth's atmosphere, causing the greenhouse effect.

Mounds of Moss

If you could shrink down to ant-size and take a trip through some moss, you would see mosses shaped like palm trees, feathers, ferns and even giant pincushions. Although shrinking is impossible, you can use a magnifying glass to get a close-up look at mosses and see some of their beautiful shapes, sizes and colors. You'll find mosses growing on rocks, tree bark, rotting logs and sidewalk cracks, even under the snow. Collect a few samples of mosses and plant a moss garden at home. Then create a rainstorm and discover the important roles mosses play in the environment.

1. Spread 2 cm (3/4 inch) of gravel in the bottom of your container for drainage. Add a thin layer of charcoal pieces to keep the soil fresh. If you are using a jar, turn it on its side and tape two sticks, a few centimetres (about an inch) apart, along one side. This will stop the jar from rolling.

2. Cover the charcoal with 2 cm (3/4 inch) of loose soil. Make the soil higher towards the back of the container so that your plants can grow at different heights.

3. Collect a few different kinds of mosses. Bring back some of the soil, bark, rock or other surface on which they are growing so you don't disturb the rhizoids (root-like parts) too much. Put your mosses in plastic bags tied with twist ties. Take a few extra samples to use in the investigations on the following pages.

4. Arrange the mosses in your container next to each other, but not on top of each other. Place the tallest mosses at the back, and add some moss-covered stones or bark for interest.

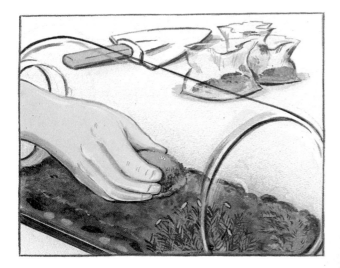

5. Sprinkle the soil lightly with water until it is damp, but not soggy. Put a lid or some plastic wrap on your container and place it in a cool place, out of the sun. Leave your container in the shade for a few days to let the plants adjust to their new environment. When you see signs of growth, such as new branches or leaves, move the container to a spot with more light, but not in direct sunlight.

6. You shouldn't need to water your moss garden since the lid keeps the moisture in. Water given off by the plants runs down the sides of the jar to the soil, where it is taken up by the plants again.

7. Your moss garden can be kept for years with very little work. As the mosses spread, you'll need to cut them back and divide and remove parts of plants so they don't get overcrowded. The plants that are taken out can be replanted outdoors. When you have finished with your moss garden, return the mosses to where you found them.

What if ...

... the sides of the glass are fogged up? This means there is too much moisture inside the container. Take the lid off for a few days so some of the extra water can escape, then put the lid back on. If the container stays too wet, some of the plants may become moldy and start to rot.

... the mosses look dry? If you notice curled or brown-colored leaves, you need to add more water to the container. Drying may also be caused by too much hot sun, so try moving the container to a shadier spot.

Habitat Watch

Before you plant the mosses, take a close look at them with a magnifying glass. Try to find the different parts of a moss shown here. Mosses are non-flowering plants. Instead of having flowers that produce seeds, mosses have spore capsules that contain tiny, dust-like spores. The spores are spread by the wind and, if they land in a good growing spot, they produce new plants.

As you look at your moss garden with a magnifying glass, compare the shape, size, color and texture of the leaves of different plants. You'll discover two main kinds of mosses. Find a moss plant that is growing upright, like a miniature tree, with its spore capsules on stalks at the ends of the stems. Mosses that grow like this are called acrocarpous. If you see a moss creeping along the ground, on rock or bark, like an ivy, then you've found a pleurocarpous moss. Its spore capsules grow from branches off the main stem.

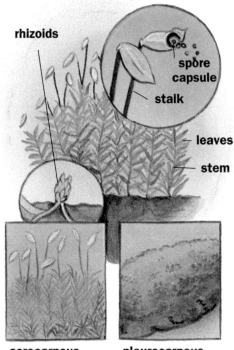

rhizoids

spore capsule

stalk

leaves

stem

acrocarpous moss

pleurocarpous moss

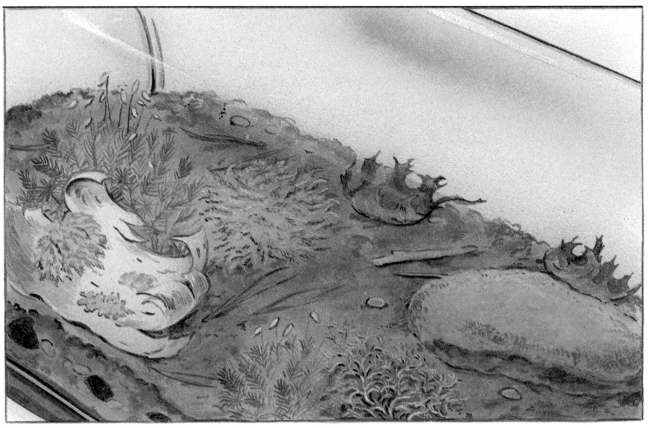

Moss Magic

What happens to your houseplants if you forget to water them? Most plants dry up and die if they don't get water. Moss dries up, too, but it doesn't die. Just add some water to some dry, dead-looking moss and it will turn back into a green, healthy plant again, right before your eyes. With a few materials, you can try this moss magic.

You'll need:
some moss
a paper towel
a magnifying glass
a bowl
water

1. Place a piece of moss on a paper towel and lay it in the sun for several days until it is dried out.

2. Compare the dried moss to the damp moss in your moss garden. How does the dried moss feel? Look at the leaves with a magnifying glass. How has the dried moss changed?

3. Place the dried moss into a bowl of water and watch what happens. Can you see the stems and leaves straighten out as they absorb the water? What happens to the moss's color? As the moss returns to its original shape, its green color comes back, too. This is what happens to wild mosses when there is a dry spell followed by rain. Instead of dying from lack of water, mosses simply shrivel up. When it rains again, they come back to life and continue growing.

Soil Saver

Plants stop the rain from hitting bare soil the same way an umbrella keeps the rain from hitting your head. When rain strikes bare ground, it can pick up bits of earth and carry them away into nearby streams or lakes. This loss of soil is called erosion. Plants that grow over the ground, like mosses do, shelter the soil from direct rain and help stop erosion. You can create a mini rainstorm in a jar and see how moss works to keep soil in its habitat where it belongs.

You'll need:
2 large plastic funnels
2 large wide-mouth jars
some soil
some fresh moss
a measuring cup
water

1. Place a funnel in each jar. Pack each funnel with soil until it is half full.

2. Cover the soil in one funnel with fresh moss. Leave the soil in the other funnel bare.

3. Pour a cup of water into each funnel to represent a heavy rainfall. Watch how quickly the water drips through the bottom of the funnel into each jar below. Compare the color of water in each jar. Where does the brownish color come from? As the water travels through the soil, it will pick up bits of soil. The water below the funnel of bare soil should be darker brown since more soil will be eroded by the water. The moss-covered soil is protected from the water so less soil is washed away.

4. Which jar contains the most water? The moss soaks up some of the water (or rain) so less water will fall into the jar below. Since moss can absorb water like a sponge does, it helps to keep the soil moist between rainfalls when bare ground often dries out. Moss also helps to keep soil in its place where it is needed by other plants.

Mosses are a kind of pioneer plant. They are among the first plants to grow in an area of bare ground or rock. The cushiony mosses provide a great landing pad for the seeds of other plants such as grasses, wildflowers and shrubs. The mosses give these seeds a place to start growing. Over hundreds of years, a bare area can turn into a forest, thanks to moss. Mosses can even create soil out of rock. The rhizoids of mosses push their way into small cracks and split off tiny pieces of rock. Special liquids produced by the rhizoids eventually convert the rock chips into soil.

Mosses also create soil when they die. As the mosses grow, parts of their leaves and stems die and decay, forming small amounts of soil where other plants can take root and grow. And you've seen how moss prevents soil from being washed away by rain. This helps keep the habitat of plants and soil animals safer.

Mosses helped renew the environment after the glaciers left thousands of years ago, and mosses are still important as plant pioneers. For example, volcanic eruptions can destroy the plant life for hundreds of kilometres (miles) around, but thanks to mosses and other plant pioneers, the barren land will become green again.

Turn a Pond into a Forest

What do you think would happen if the grass in your yard or local park was never cut again? At first you'd be wading through a lot of tall grass. Gradually, the seeds of nearby plants would be carried by the wind and settle in the grass. Soon, you'd have lots of plants growing along with the grass and, after many years, it would look like a small forest. This slow replacement of one kind of habitat (your yard or park) by another habitat (a forest) is called succession. Succession is going on all the time; it's part of nature. Succession turns abandoned farmland into forests, changes small ponds into dry land and helps areas burnt by fires to become green again. Instead of waiting 20 years to see succession in action, you can set up a mini habitat at home and watch a pond turn into a forest.

You'll need:
gravel
a plastic dishpan about
  30 cm x 35 cm x 12 cm
  (12 inches x 14 inches x
  5 inches)
potting soil
a plastic plant saucer 3 to
  4 cm (1 to 1 1/2 inches)
  deep and 10 to 15 cm
  (4 to 6 inches) wide
water
grass seed
mixed bird seed (including
  sunflower seeds)

1. Place 3 cm (1 inch) of gravel in the bottom of your dishpan for drainage. Cover this with about 10 cm (4 inches) of soil.

2. Sink a plant saucer into the soil in the center of your dishpan, so that the top edge is level with the soil surface. Put about 1 cm (1/2 inch) of soil in the bottom of the saucer. This will become your pond.

3. Slowly pour water into the dishpan so that the pond is full of water and the rest of the soil is wet. Place the uncovered dishpan on a table in a sunny window.

4. Sprinkle a handful of grass seed over the entire dishpan, including the pond. After a few days, check to see what is happening. Do you see any green shoots poking out of the earth? What is happening to the seeds that landed in the water? The seeds that were sprinkled on the earth will start to grow after a few days. Those that landed in the water may sprout, but they won't grow because they cannot get any air. Eventually, the water-logged seeds will rot. Leave the rotten seeds in the pond. They represent the natural build-up of dead materials on a pond bottom.

5. Continue sprinkling a handful of grass seed over your dishpan every 3 or 4 days. You will notice that the pond is becoming shallower. This is partly because some of the water is evaporating and partly because of the build-up of material on the pond bottom. Eventually, your pond will dry up. Natural shallow ponds often dry up during hot summers when there is very little rainfall. Also, decaying material from aquatic plants and materials blown into the pond settles on the bottom of the pond. Gradually, the layer of litter builds up, making the pond shallower. Eventually, the pond is shallow enough to allow non-aquatic plant seeds to sprout and grow.

6. Lightly water the soil in your dishpan to represent rainfall, but do not refill the pond. What happens to the grass seed that falls into the empty pond? Now that the deep water is gone, the grass should start to grow in the pond basin.

7. You should find that an open field or meadow of grass has replaced the bare ground and pond that you started with. Now, instead of grass seed, sprinkle a handful of mixed bird seed over the whole dishpan once a week for two weeks. The plants that grow from the bird seed — such as sunflowers and millet — will be much bigger than the short grass plants. These new plants represent the gradual invasion of shrubs and trees into the field habitat. Eventually, you will have a mini forest of large plants.

8. When you are finished with your mini habitat, you can transplant your sunflowers into a garden or large pots outdoors.

When you cut your grass, you cut everything. This prevents natural succession from taking over your lawn and eventually turning it into a field or forest. Succession can also be interrupted by nature itself. For example, imagine a pond that is gradually being filled in naturally and its water is just getting shallow enough for land plants to start growing along the mucky shore. If a flood occurred, the pond level would rise, the shoreline plants would be water-logged and die, and the pond would be back to where it started. On the other hand, if there were a drought, the pond would dry up faster and land plants would be able to invade sooner, speeding up the natural process of succession. A forest fire, earthquake, tornado, volcanic eruption, drought or flood can all destroy habitats and create new ones.

Have you ever seen a fish in a tree or a woodpecker in a pond? Of course not. Each kind of animal is designed to live in its own

special habitat, and when the habitat changes, so do the animals that live there. As a pond dries up, the aquatic plants and animals can't live there any more. Land plants, such as wildflowers and grasses, move in, along with land-loving insects and other tiny organisms, small mammals such as mice, and small birds. Over time, if the open field becomes a shady forest, some animals leave and new ones move in. Woodpeckers, wood-boring insects, squirrels and other tree-loving species will make the new forest their home.

As the population of our world rises, we take over more land to support people. This means that more natural habitats are destroyed. Wetlands are drained for farming, dredged for marinas and filled in for construction; forests are cleared for roads; and prairie grasslands are plowed under for crops every day. Once the natural habitats are gone, they are lost forever. Over time, new habitats may grow to replace the old ones, but the new growth will never be exactly the same as what was lost.

We need to save some of the remaining examples of different kinds of habitats around the world. An international group of scientists and environmentalists is creating a list of areas that governments should protect, such as grasslands, wetlands and rainforests. To make sure that these habitats are preserved for future generations and for the future health of our planet, countries are setting up wilderness zones, parks and other special areas where development is not allowed. Anyone who owns property that has special or rare habitats on it should protect that land and keep it as natural as possible. Here are some ways that you can help:

• Observe or take pictures of wildflowers in their natural setting instead of picking them.
• Ask your parents to leave some of your property uncut.
• Suggest alternatives to chemical sprays for weeds and insects.
• Keep nature litter-free.
• Set up a Neighborhood Habitat Watch club where members keep a look-out for local habitat destruction. Tell a local Conservation Officer about any problems you discover. Nature needs your help.

Index